T0224339

Sustainable Innovation

A Guide to Harvesting the Untapped Riches of Opposition, Unlikely Combinations, and a Plan B

Sustainable Innovation: A Guide to Harvesting the Untapped Riches of Opposition, Unlikely Combinations, and a Plan B
Lisbeth Borbye

ISBN: 978-3-031-01445-1   paperback
ISBN: 978-3-031-02573-0   ebook

DOI 10.1007/978-3-031-02573-0

A Publication in the Springer series
SYNTHESIS LECTURES ON TECHNOLOGY, MANAGEMENT, AND ENTREPRENEURSHIP

Lecture #3
Series Editor: Henry E. Riggs, *Founding President and Trustee Emeritus, Keck Graduate Institute*
Series ISSN
Synthesis Lectures on Technology, Management, and Entrepreneurship
Print 1933-978X   Electronic 1933-9798

Disclaimer: *The content of this book is based on the author's personal ideas, knowledge, and experience. The author disclaims any liability or loss in connection with use or distribution of the information herein. Use of the information is entirely at own risk. The contents of this publication do not constitute legal or professional advice. Readers should not act or rely on any information in this book without first seeking the advice of relevant experts.*

Synthesis Lectures on Technology, Management, and Entrepreneurship

Editor

Henry E. Riggs, *Founding President and Trustee Emeritus, Keck Graduate Institute*

The titles in this series permit readers to select topics that will better equip them for improved performance and advancement in enterprises, public and private, pursuing strategies that leverage technology – including information processing, bio- and nano-technology and other rapidly advancing fields – to create, produce and market innovative products and services.
Among the topics to be addressed by this series are:

- Leadership in innovative organizations

- Accelerating innovation

- Deciphering essential financial information

- Project management

- Marketing technology-based products and services

- Inspiring, coaching and guiding the creative process

- What does entrepreneurship mean?

- The challenge of multi-disciplinary teams

- High-performance organizations

Sustainable Innovation: A Guide to Harvesting the Untapped Riches of Opposition, Unlikely Combinations, and a Plan B
Lisbeth Borbye
2011

Out of the Comfort Zone: New Ways to Teach, Learn, and Assess Essential Professional Skill—*An Advancement in Educational Innovation*
Lisbeth Borbye
2010

iv

Understanding the Financial Score
Henry E. Riggs
2007

Sustainable Innovation

A Guide to Harvesting the Untapped Riches of Opposition,
Unlikely Combinations, and a Plan B

Lisbeth Borbye
North Carolina State University

*SYNTHESIS LECTURES ON TECHNOLOGY, MANAGEMENT, AND
ENTREPRENEURSHIP #3*

ABSTRACT

Many of us wish we could design inventions and make decisions that were optimal and sustainable, but we do not know how to begin the approach. This book offers a guide to dramatically improve the quality of innovation and solution-making through the respectful use of existing and abundant, but often-ignored, resources. Sustainable innovation is about creative combination of ideas, materials, methods, and people, courage to derive value from opposition and diversity, integrative intelligence, virtuous planning, minimal consumption of resources, and definition of alternative plans. Using the method successfully requires that we are truly interested in the common good of humankind, that we care about our environment, and that we take time to think carefully about consequences before we act, invent, or make decisions. It is a call for a much-needed collaboration between people of different backgrounds, skills, and opinions with the intent to preserve individual and local integrity and adapt a win-win mindset. Try it, and partake in its positive and long-lasting effects!

Key Features

- Description of sustainable innovation and the untapped innovation potential

- Sustainable innovation requirements

- Sustainable innovation attitudes

- Step-by-step guide to sustainable innovation

- Application to 21st century challenges

- A global win-win scenario

KEYWORDS

sustainable innovation, integrative intelligence, unlikely combinations, diversity, opposition, peaceful solution-making

In admiration of all who invent and provide solutions
for the sake of the common good

Contents

Preface xi

1 Innovation Attitudes 1
    1.1   "Best in the world" 1
    1.2   Current innovation incentives and barriers 1
    1.3   The rise of severe crises: too big and quick to not fail 2
    1.4   Mindless innovation and consumption 3
    1.5   Defining the need 3
    1.6   Towards sustainable innovation 4

2 Inspiration from the Natural and Behavioral Sciences 7
    2.1   "Extreme innovation" 7
    2.2   Natural superior innovation 7
    2.3   Synthetic superior innovation 8
    2.4   Superior behavioral scenarios and the "Plurosis" principle 8

3 Untapped Sustainable Innovation Potential 11
    3.1   Sustainability 11
    3.2   Responsibility 11
    3.3   Superiority 12
    3.4   Peace 12
    3.5   Before you begin: sustainable innovation requirements checklist 13

4 Step-by-Step Guide to Sustainable Innovation 15
    4.1   A word about interaction dynamics 15
    4.2   Before you begin: sustainable innovation attitudes checklist 16
    4.3   Overview of Steps I - IX 17
    4.4   Steps I – IX: Questions Section 19

5 Notes on the Application to 21st Century Challenges 27
    5.1   Sustainable energy production and consumption (Steps I – III) 27

5.2   Environmentally-friendly, abundant food production (Steps III – V) 28

5.3   Control of emerging diseases (Steps VI – VIII)............................. 30

5.4   Verbal disputes and constructive freedom of speech (Steps IV – V, IX) 31

6   Personal Step-by-Step Work Section 33

7   Looking to the Future 53

7.1   Impact of sustainable innovation 53

7.2   Consequences of global competition 54

7.3   A global win-win scenario 54

Inspirational Reading 57

A   Notes on Teaching Sustainable Innovation 59

B   Examples of Unlikely Combinations Parameters 61

C   Sustainable Innovation Requirements Checklist 63

D   Sustainable Innovation Attitudes Checklist 65

E   Examples of Drivers and Barriers to Sustainable Innovation 67

Other Books by the Author 69

     Secrets to Success in Industry Careers: Essential Skills for Science and Business. 69

     Industry Immersion Learning: Real-Life Case Studies in Biotechnology and
Business.. 71

     Out of the Comfort Zone: New Ways to Teach, Learn, and Assess Essential
Professional Skills. 73

Author's Biography 75

Preface

We must continuously examine the intent behind and process with which innovation takes place to ensure that it is aligned with our value systems. The "too big or too quick to not fail" phenomenon presented in this book serves to inspire a renewed focus on integrity, strength in diversity and improved planning and quality control practices. With a win-win mindset, I set out to shed light on the questions:

1. *"How do we characterize, develop, and implement the best possible inventions and solutions?"*

2. *"Are we brave and free enough to dismantle or improve existing, subpar solutions?"*

The great news is that we have abundant resources at our disposal. Application of self control, gratitude, courage, and attitude mastery help extract the "gold" from opposing viewpoints. Integrative intelligence helps define the congruencies between different opinions and ideas to synthesize the best of all worlds and anticipatory, visionary problem-solving to develop contingency plans promises a triple bottom-line of responsible, superior, and peaceful innovation.

Let's get started as soon as possible.

Lisbeth Borbye
May, 2011

CHAPTER 1

Innovation Attitudes

1.1 "BEST IN THE WORLD"

You have probably often heard the phrase "best in the world" and wondered whether the person saying it knows exactly what is best and has seen the entire world and therefore is a trustworthy source on the subject. Alternatively, it is possible that the person talking is not thinking about the same world as you. Therefore, claiming to be or have the very best in the world is a relative business and depends on the definition of "the best" and "the world" by multiple and complex parameters. For example, "world" can be defined using natural geographical boundaries that create element and climate zones (ocean vs. land, tropics vs. arctic, etc.) or by man-made artificial divisions making up countries and cultures, and even language). Therefore, what is "best" depends on the individual and his or her situation. This means that a wealth of circumstances ranging from psychological heritage, personal preferences, age, ideological or religious conviction, to education and experience levels play a role in the quality judgment of an item, idea, location, or process.

The bottom-line is that we humans like to strive to "be and do the best," a quest that is so prevalent in society that it can lead to abandonment of less-than-optimal strategies when better ones have been identified. The challenge is to understand that the thoughtful consideration and meshing of multiple ideas is "best" because what we do has consequences for many. Ultimately, we serve ourselves and others best when inventions and solutions are sustainable, defined by being responsible, superior, and peaceful, and can therefore stand the test of time.

1.2 CURRENT INNOVATION INCENTIVES AND BARRIERS

People innovate for a number of reasons. Many will respond to basic human needs, such as the urgent need for pure drinking water or shelter from the sun. Other reasons for innovation include monetary incentives, idealism, enthusiasm, pleasure, and play. Certain personalities thrive on conceptualization and others from idealism and innovation. Some simply enjoy pushing boundaries and exploring the hitherto unexplored.

An invention is defined as something that is new, slightly altered, or changed. If it is imagined (or in some cases, such as the sequence of a gene, discovered) by the human mind, there are ways of claiming possession of the invention. In other words, the invention may be classified as intellectual property and be patentable if it fulfills certain legal criteria. These criteria relate to the operations, utility, novelty, non-obviousness, and enabling description of the invention [1].

Most businesses have commercial incentives to build substantial intellectual property inventories; that is, they patent inventions and sell or license their products and ideas. Alternatively, they may choose to hide the inventions from others in order to preserve their own knowledge-based advantage. For this to work properly, there must be a set of common agreements and principles to which anybody that has access to the intellectual property must adhere. It is well-known that not all inhabitants of our Earth have the same understanding of how to comply with these agreements. The differences are often a source for accusation of illegal activities such as simple plagiarism (for example, violation of copyright), falsification of designer labels, or copy-cat mass production of generic products. This situation increases protectionism and a reluctance to share information because the livelihood of many businesses depends on developing and protecting intellectual property. Likewise, if the cost of improvement seems too high or if the business is already profitable, incentives for new innovation may be dampened.

The current need for sustainable innovation and solution-making methods reflects that the Earth is home to all kinds of people. Such diversity seems ideal for innovation, but at the same time, innovation is often hampered by ethnic, spiritual, ideological or practical differences, or a lack of collaborative communication, monetary and structural resources, and a logical process. Instead, we tend to do things in groups of lesser diversity, within a well-defined business structure with little or no understanding of the implication of and for the parties that will likely be affected by what we do (including the environment) and without considering alternative solutions.

1.3   THE RISE OF SEVERE CRISES: TOO BIG AND QUICK TO NOT FAIL

We continuously face challenges and crises in the world, some of which affect a large number or population of people. Examples include natural disasters, global pollution, mine and oil well accidents, nuclear plant melt-downs, revolutions, and economic distress. Often, it seems like there is no obvious solution or functional backup plan and human-created system failure caused by priority and excitement for economic gains and large-scale dependency have become a common theme. These are features that tend to minimize the much-needed scrutiny and quality control of both process and product. The financial crisis, which started in 2008, illuminated a need for responsible and transparent measures and regulations of both Wall Street's trading practices and commercial lending. It is likely that additional crises are only the beginning of several disasters that we will need to avoid or manage because things have evolved too rapidly to be of good quality.

In addition to high-speed development, sheer size is a risk because it drives monopoly. The more people who are dependent on a service or product, the more havoc will the absence of the service or product create, due to the presence of limitations and scarcity if no back-up plan (or exchangeable product) is in place. In addition, large enterprises may become hide-outs for system errors and risky, unethical behavior. Many of our current provision systems such as energy and food supplies, e-information technology, financial services, transportation, and many more fall in this fragile "too big" category.

1.4   MINDLESS INNOVATION AND CONSUMPTION

A striking feature in cultures blessed with a period of wealth is the abundant offering of unnecessary products (defined as products one can live happily without). Many can relate to the excess of different kinds of cereals, plastic toys, electronic gadgets, and games. While it is worthwhile to celebrate that these items all are products of creative minds and to remind ourselves that profit is still essential for continued production, it is timely to ask whether it is sustainable and reasonable to create something and produce it, to buy and consume it, just because we can. The same goes for consumption of resources for research "pet projects," unnecessary meetings and procedures, or any possible creation of "tasks for tasks sake" (providing so-called "job security"). A by-product of mindless innovation and excess consumption is unnecessary waste. Anything that creates waste (of time, resources, as a product after use, or unnecessary storage) is, or contributes to, a disaster in the making.

Why should we be concerned about this topic? The problem is that we are lowering the standard of living (by creating waste, pollution, and disparity) for current and future generations with our present way of life. Furthermore, the incentives for greed, being in charge, and being "number one" are becoming more challenging as population density increases. It is not difficult to imagine additional unrest, malignant competition, and a rise in violence as a result.

1.5   DEFINING THE NEED

One way to put an end to mindless innovation and consumption is to think about what WE REALLY NEED as well as to prioritize what we need to do and how quickly we should do it. To answer this, we will also need to focus on not starting, not consuming, or eliminating what we do not need. Production and consumption must be responsible, effective, sustainable, and nurture peace and prosperity in terms of providing the necessary food and water, clothing, shelter, safety, healing, and care for the people of this world. Once this is done, there may be room for a few luxury gadgets! This task also includes estimating whether or not "time is ripe" for a certain invention/solution. If the time is not ripe, much well-intended effort most likely will not create the wanted outcome but rather delay the process. To prepare for "ripeness," it is helpful to address all or most of the concerns offered by opponents prior to the introduction of a invention/solution.

Although need is relative to the developmental and situational stage of a culture, a sole focus on the individual inventor's profit is problematic because inventors are expending resources on behalf of the Earth's entire population to reach their goals. Also, to define the collaborative need, we must work together to gather data and to provide inventions and solutions. This includes sharing information and ideas. Perhaps we may even play with the idea that each person on Earth has a consumption allowance, for example defined by the "carbon footprint" [2], and hydrogen and oxygen accounts. For this to work, the range should reflect what would seem sustainable within the given culture and should be wide enough to provide incentives for individuals to work happily towards their "dreams coming true" as well as to create opportunities for innovation. Another controversial idea to control consumption is population management. Maintaining a certain population size simply implies that each person can reproduce once, i.e., a couple would have a maximum of two children. An alternative approach to controlling the "carbon footprint" and population growth is to maintain or increase a "positive ecological footprint" by using, and re-using resources in a cyclical fashion as opposed to creating goods without an end-of-current-use plan in place [2, 17].

We know there is a long cadre of both current and upcoming predicted challenges and some examples are given here (and further explored in Chapter 5):

1. Our current energy supply and the supply methods are insufficient and rely on few and limited resources.

2. We do not know how to best adapt to climate changes and develop alternative solutions to avoid aggravating the situation.

3. We have developed unhealthy lifestyles, which make us more susceptible to current and emerging diseases and degenerations.

4. New diseases are expected to emerge with globalization and overall increased mobility of man; these are expected to spread quicker than before.

5. The way we mass produce food makes animals and crops more prone to pest damage and increases the risk for famine.

6. The world population is increasing, which places larger demands on the food supply chain and creates an even higher risk for pollution.

7. The ease with which and level [3] at which we can learn, communicate [4], and travel on a global scale, in combination with the differences between us, is causing polarization and a rise in both extremist behavior and a lack of a common code of conduct.

8. The "win-lose" thinking preference is dominant and causes a high level of waste of resources.

1.6   TOWARDS SUSTAINABLE INNOVATION

Recall a heated conversation between family members, an employer and an employee, or between politicians or countries of different political orientations. It is clear that we are passionate about what is "right and best" in this world, and it is also clear that most of us think that we know exactly what this "right and best" is. But, it is unlikely that one single human being would be so enlightened as to know the correct answer to everything. Even if such a person existed, history has revealed that the rest of us probably would not agree. That is why it is so important to learn the benefits of disagreeing peacefully or not to agree on everything instantly and instead take time to debate and weigh pros and cons. When people oppose a view, they are usually conveying an important message of concern, a treasure of information. Therefore, we must learn to muster curiosity and an understanding of the necessity to peacefully respect different views [5] and personas [6] as these contain enormous value and opportunity.

A fascinating aspect about these challenges is that they constitute an untapped innovation potential and if we learn how to communicate, acknowledge, and collaboratively use the cumulative potential, the superiority of inventions and solutions promise to be sustainable and much greater than those created by solo enterprises. A simple example is the inclusion of people with a creative preference (so-called right-brainers, for example, artists) on teams with a typical "left brain" orientation (for example, scientists, mathematicians) [7] or the acceptance of the "odd-ball" employee, who in return may contribute a great new idea [8]. The goal is not to achieve individual victory, but to mold the best collective solutions after taking into account multiple opinions and concerns. Sometimes, this may mean having a subset of different solutions in different cultures as part of an overall strategy. So, contrary to popular belief, we may learn to welcome diversity and opposing views, think more freely, and find ways to integrate troubling extremes, while increasing our tolerance towards others and their values.

Cultures and people who can afford to create inventions benefit from including humanitarian purposes in all of what they do because they become positive role models to themselves and others. So-called "social entrepreneurship" [2] and "life entrepreneurs" [9] provide such an example. Life entrepreneurs include a social purpose in their business plans, and a reason for doing so is often a wish to have done something good for a "neighbor" (such as people in need). Research has shown that such openness and giving to others enhances positivity [10], a condition that strengthens awareness about beneficial aspects of being in the present moment. In earlier times, this concept was often confused with the more transient feeling of "happiness." Needless to say, social and life entrepreneurship seem like good ways to undercut, or perhaps even eradicate, extremism in current and upcoming generations and to reduce jealousy and suspicion from other people, cultures, and countries.

Switching from promoting unnecessary excess to the principle that getting rich means taking care of BOTH one's own and others' needs is making space for a much-needed sustainable life style. This implies a focus on shared values instead of instant tangible personal gratification and is important for all of humanity but also for the individual for reasons indicated above. It is not easy, mostly because we are not used to thinking this way. Fortunately, many entrepreneurs are now assisting this process by sharing ideas in interdisciplinary think-tanks and open innovation forums. So, what are the incentives to work productively together to face current and upcoming challenges and to create the best of all worlds? My hope is that understanding the advantages of optimizing our current innovation and solution-making paradigm followed by access to a logical method of doing so is sufficient, especially because the resources are abundant, free, and accessible. I describe my source of inspiration, the essential components of my version of sustainable innovation, a logical framework for harvesting the untapped riches of opposition, unlikely combinations, and a "plan B," and beneficial outcomes for future generations in the following chapters.

CHAPTER 2

Inspiration from the Natural and Behavioral Sciences

2.1 "EXTREME INNOVATION"

The concept of "extreme innovation" is a relatively new "buzz" word, that deserves an appropriate definition. Some think that it is the product of the most innovative ideas, which by others may be perceived as extremist or radical. In this chapter I will use the term "extreme innovation" to describe a combination of extremes to create superior products and solutions. Nature has its own ways to develop such products, and man has learned to understand, "copy," and "force" nature to create, what are considered to be optimal products. In particular, this is a well-known and accepted practice in the field of plant breeding. From this practice, I deduce the unusual combinations principle to pervade and inspire optimization in innovation. Because opposition also can be characterized as "extreme" opinions or polarized values, opposition itself has the potential to enrich innovation substantially.

2.2 NATURAL SUPERIOR INNOVATION

Natural "extreme" and superior innovation may be seen as an integral part of Darwin's evolution theory in the sense that it allows survival of the fittest and/or those who benefit from or create benefits for others. It is well known that certain traits provide advantages in a particular environment, whereas others do not. Examples from the plant world are resistance to pests and the ability to tolerate extreme weather, such as wind, droughts, and heavy rain. These traits are harbored by certain genes, and these genes are inherited naturally by plant progeny.

Nature has a repertoire of tools to create variation and optimize the species' conditions for survival when change is encountered. It includes events such as transfer of genes between different species and alteration of genes (for example, genes can be added, multiplied, deleted, partially deleted, activated, or deactivated). This implies that nature has its own methods for creating gene modified organisms (GMOs). Changes in natural organisms' genes can be induced by sunlight (UV radiation), by the strike of lightning, by transfer of genes by or from natural microorganisms, by exposure to chemicals (such as pesticides), and by other means.

Some of these changes are detrimental (which means the plant and/or its progeny will die) while others may create a superior plant. The plant may now be resistant to a disease to which it was susceptible before the change. It may carry more fruit and produce a higher yield. When the progeny is deemed superior to its ancestors, a special heterosis effect may be attributed. Superiority can also be induced when different plants are crossed (either spontaneously or by man (see below)). This effect can inspire our innovation practices. What is important to understand is that the introduction of unexpected or unrelated ideas, materials, methods, events, and parties can create optimal situations.

What about opposition? Even plants fight it out! Weeds compete with crops; all compete with all, and the fittest survive. Yet, in many natural habitats, mutually beneficial relationships exist in which different parties "collaborate" and gain in unusual ways from the presence of each other. A popular example is the symbiosis present in lichens in which algae and fungi help nurture each other using their differences to sustain the entire physical entity, an entity different from what either organism would have looked like separately.

2.3 SYNTHETIC SUPERIOR INNOVATION

Traditional plant breeding regimens work toward the accumulation of beneficial traits and surveillance of the occurrence of a heterosis effect as a way to create new and better crops and ornamentals. Often, traditional plant breeding, in which plants are cross-pollinated, is also looked upon as "natural," even though plants are crossed manually by man. The same is the case for manipulation techniques such as grafting a new tree onto a different mother stem.

Synthetic plant breeding practices include a variety of additional manipulation techniques such as those performed on mini-plants (in vitro culture), and on cellular and molecular (DNA) levels. The purpose of all of these techniques is to develop a superior product. One could imagine the fusion of a carrot plant and cilantro plant in order to create a fully edible plant for humans or the creation of a super-sized blueberry, so picking the berries would not be as laborious, etc. These are examples of a synthetic (or forced) beneficial union of different traits. My point here is not to fully describe or judge these techniques but to draw inspiration.

2.4 SUPERIOR BEHAVIORAL SCENARIOS AND THE "PLUROSIS" PRINCIPLE

Heterosis implies that we arrive at a superior product by virtue of merging two different entities. Parallels may be drawn to peacemakers, who resolve conflicts by merging different and often opposing views. Marriage is another example. If marriage is to be superior to individual habitation, the effect of being together must create a new, attractive entity of additional value to both individuals when compared with the value of the individuals' separate lives. Imagine applying this principle to multiple materials, methods, levels, opinions, people, and situations to derive an outcome, which I call "plurosis." Plurosis is defined here as the superior outcome created by merging a variety of elements and factors that lead to unanticipated and often unexplainable benefits. It can be interpreted as "optimal inclusivism."

We can request input from very different people to create the conditions necessary for "plurosis." Differences can be defined in terms of age, sex, ethnicity, country of origin, culture, ideological, religious and spiritual orientation, education level or field, other fields of expertise, marital status, level of global understanding, "life experiences" (such as child birth, marriage, divorce, injury, disease, death, etc.), social economic standard, personality type, and many others. These differences will ensure that unusual combinations, opposing views, and even fears are present. Employing a melting pot "plurosis principle" is essential to the quest for sustainable innovation.

CHAPTER 3

Untapped Sustainable Innovation Potential

3.1 SUSTAINABILITY

Elevating innovation to become sustainable will most likely require doing things in new ways. We may have to move away from our comfort zones into uncharted territory to deal with the unfamiliar and uncomfortable to potentially gain the highly beneficial [11]. The desires for a meaningful life and a purposeful occupation [12] will hopefully lead us on the path towards sustainable innovation.

What does sustainable innovation mean? In brief, it means innovation can last for a long, long time while having a positive impact on all it affects. Such high standards can be achieved by requiring that innovation be responsible, superior, and peaceful. Knowledge and technology must be or become available to provide outcomes that match the ("lofty") end goal. The inclusion of high humanitarian and ethical standards and well-thought-out solutions are at the pinnacle when we attempt to create the very best. Individual cultures must work out their own measures in each of these areas as these may differ until we have acquired a mindset as United Citizens of the Earth (which is expected to take a while). At the same time, globalization requires that we are able to work in multi-cultural teams and master skills for successful interaction. Such skills include cross-cultural understanding, flexibility, ethical conduct, active listening and other communication and inter-relational skills [13]. We need to nurture a mindset of positivity [10] and belief that we will be able to achieve what was earlier not possible or barely envisioned, since this belief may strengthen the likelihood of a wishful outcome [14].

3.2 RESPONSIBILITY

An invention/solution must meet an essential need (other than personal monetary gains) to be deemed responsible, and it must be aligned with a generally accepted and honorable code of conduct. Such a code speaks to the best in each of us and includes truthfulness, trustworthiness, and an interest in the common good. Responsible innovation is surrounded by the best intentions for all parties that will be affected. The "creators" are willingly accountable, and their intent is to improve a situation.

The positive impact of innovation must be assessed as substantial, while the negative impact must be acceptable. This relates to multiple and highly-variable areas, such as the consumption of resources and creation of waste (during innovation and post-innovation usage), understanding of change to both beneficiaries and adversaries, and the environment. Most importantly, responsibility includes knowing what happens and what to do if an invention fails. Will failure create a crisis? If yes, one or more back-up plans must be in place. If the crisis is due to dependency, an emergency substitute product and/or process needs to be available. If the crisis is caused by pervasiveness (interdependent global units or size), a back-up plan could consist of the immediate dispersal into smaller self-sustainable and independent units, similar to the isolation of a fire by emergency doors.

3.3 SUPERIORITY

Superior innovation employs the potential effect of inclusion or subtraction of different components. These components may relate to a number of essential, structural or other differences such as product specifics (origin, composition, design, etc.), viability measures (productivity, longevity, etc.), and indication or usage.

One must be very knowledgeable about individual components and what has already been accomplished, or courageously attempt to combine new matters (ideas, materials, and methods), new matters with old matters, or old matters in new or unusual ways to create something superior. Alternatively, both serendipity and luck have been the basis for many inventions. In all cases, measures for superiority need to be established. Synergy can also be found in human interactions. Therefore, using attributes such as an uncommon combination of people is expected to greatly strengthen innovation. Examples of often underappreciated resources include the honesty of children and the wisdom of old people.

3.4 PEACE

Peace depends on agreement among most or many. This means that most or many must be represented and implies a high level of population diversity. Peaceful solutions are easy to achieve when there is agreement, but opposition carries great potential for creating lasting peace when seen as a tool for improvement and used in conjunction with active listening and conflict resolution techniques. This has to do with the enhanced involvement and responsibility solution-makers feel towards their "product." If we understand why there is opposition, it may become possible to mitigate it. Therefore, sustainable innovation depends on collective, collaborative intelligence, and openness towards criticism and conflict.

Most of us have a tendency to think our view is the correct one, and when faced with criticism we are quick to deflect it as jealousy or negativity. The truth is there often is a "gold mine" hidden in criticism and different points of view, and if we dare listen to it carefully, our views and ideas may become refined and more suitable than before. This is especially true when we feel uncomfortable and our emotions become engaged. Learning to work with opposition is like learning to walk in somebody else's shoes, even when they do not seem to fit. It requires emotional intelligence [15]. Listening and attempting to find "common ground" for most is not only about establishing trust in unity and innovation itself, it is about the advantage of simultaneously gaining access to multiple neuronal networks instead of relying solely on our own. Learning about and practicing compassionate and nonviolent communication can assist this process [16].

3.5   BEFORE YOU BEGIN: SUSTAINABLE INNOVATION REQUIREMENTS CHECKLIST

In summary, to create responsible and peaceful inventions and solutions, the inventor/invention/intentions/benchmarks must address the following:

1. *NEED: Meet essential need(s) or improve a situation, and provide substantial positive impact to beneficiaries.* Parties must agree on what the term positive impact means and identify how it can be measured.

2. *VIRTUOUS CONDUCT: Be honorable (as defined by actions to be proud of based on an established common ethical code), truthful (telling the absolute truth and providing the necessary details), trustworthy (upholding the truth and ethical code of conduct), and show interest in the common good* defined by a positive impact on the majority of affected parties.

3. *ACCOUNTABILITY: Assign personal accountability and display best intentions for affected parties.* Inventors/solution-makers must be identifiable and held responsible for their actions/inventions. The intentions should be clearly expressed.

4. *CONSUMPTION: Create reasonable, acceptable consumption, produce only acceptable, re-usable, or no waste, and cause acceptable impact on the environment.* The parameters for how many resources can be used to create a certain invention/solution and how much waste it will likely create should be established. The parameters also include how much and which kinds of waste can be tolerated, if any.

5. *ALTERNATIVES: Include one or more back-up plans (substitute innovation or breakdown into independent units).* Establishing these plans makes us think about the end-products' flexibility. We must practice anticipating the future best and worst case scenarios, which can contribute to substantial value-adding and avoid or mitigate unexpected disasters.

6. *SUPERIORITY: Have clear measures of superiority.* Knowing what it means to create a superior solution includes knowing how to measure whether it was achieved. This includes establishing tangible measures of success.

7. *UNLIKELY COMBINATIONS: Use unusual combinations of ideas, materials, methods or people, or provide enrichment by including opposition and concerns.* Providing evidence for a mutual, heterosis, or "plurosis" effect is important. Without taking this effect into account, it is very likely that another, more superior innovation is easily conceivable.

8. *COMPROMISE: Be able to find "common ground."* A prerequisite for inclusion of different ideas, materials, methods, opinions, or people is that flexibility will be displayed and/or compromise will take place.

CHAPTER 4

Step-by-Step Guide to Sustainable Innovation

4.1 A WORD ABOUT INTERACTION DYNAMICS

Launching a new idea for an audience can be a daunting task. The process of exciting people can be even more daunting. That is why it is important to be patient and helpful to gain some basic knowledge about communication and the way we perceive and react to it. The behavioral sciences and business management fields both provide clues about how we may categorize steps in the "communication cycle." For example, when starting a new team project, team members may experience an initial time of getting to know each other and the project parameters, followed by vivacious (and perhaps deeply emotional) discussions (and turf battles), a phase of reaching the middle ground, and finally finding the way to high performance.

New ideas may be met with "shock" or distrust, denial of need, and opposition. There is no need to be discouraged because a strong opinion is often a better indicator for opportunity than no opinion. At the same time, there may be ways to avoid a total "denial" or contrariety. Being of an entrepreneurial mind, I am used to giving people the benefit of time, and I often preface my presentations with the phrase "something to think about." This empowers the listener to make his or her own judgments about the suggestions. My observation is that time may even provide a distribution of ownership of the idea. In other words, if I hear my idea from somebody else after a while, then I know time has had a ripening effect, and the phase of acceptance and exploration (which is what is needed for innovation) has arrived.

It is worth having in mind that different personalities contribute value to an interaction. As an example, a person who is very data-oriented and likes to explore may introduce the newest scientific findings. A person who worries a lot or a person who relies on intuition may be able to anticipate the best and worst future scenarios and come up with a good "plan B." Another person may be extremely enthusiastic and people-oriented and enhance the social entrepreneur aspect, inviting new people to join a discussion. An optimal interaction can occur when each person understands that differences are assets for a combinatorial solution rather than obstacles to enforcing one's own opinion.

4.2   BEFORE YOU BEGIN: SUSTAINABLE INNOVATION ATTITUDES CHECKLIST

We are all together on Earth for a limited amount of time. We have this little time to practice caring for each other and the planet. This can be hard because we have expectations about doing well and striving to be the best, which implies that somebody else is not doing as good. This usually leads to competition and a win-lose scenario (which, in the long term, is a losing scenario as described in Chapter 7). Cultivating sustainable innovation attitudes can help heighten a collaborative expectation and general well-being; this is essential for an effective win-win outcome. These attitudes are a result of courage to be inclusive and open, to respect and hold humility for all human beings, and to show honesty about and acceptance of the most optimal solution regardless of whether or not it is a compromise. Essential attitudes for sustainable innovation are described below:

1. *INTENT AND CLARITY: Agree that there is something to improve/invent/resolve, clearly define it, and strive for a invention/solution.* Setting a goal of what to achieve is instrumental to success. Without this goal, it is unclear what needs to be done, and the likelihood of a change is low. It is important to want to strive to meet the goal. Some conflicting attitudes are those in which you A) do not care about whether or not a invention/solution will be found and B) do not want a invention/solution for any number of reasons.

2. *FLEXIBILITY AND CURIOSITY: Be flexible, display openness, assume that others' input is important for the end solution, and that you do not have all the answers yourself. Understand that society is dynamic and that you need to be flexible too.* When discussing and including other peoples' opinions and ideas, the invention/solution is strengthened substantially. Being able to "borrow" a different set of brains is a gift, and being able to borrow many simultaneously could create a revelation! Try not to be intimidated by criticism or great additions to your own ideas; instead find the value. Society is likely to change faster than you do, and to keep up, you will need the help of others.

3. *FORGIVENESS AND RECONCILIATION: Avoid focusing on what has been fair and unfair; instead focus on a new best and just outcome. Avoid blaming anybody; instead display a forgiving or reconciliatory mindset. Use active listening skills [13].* What has happened in the past can easily create preconceptions regarding the future. Much has likely been less than optimal, but it is not going to help to keep rehashing it. Without naïveté, try to look objectively at inventions/solutions and dare to think the best. A conversation in which people are made to feel guilty is rarely productive. There are other ways to address issues that you do not want to have happen again. Say exactly what outcomes are desired and what you want from people rather than what you do not want or what others have done wrong. Also, when you are able to forgive what has happened in the past, you may feel free to pursue a new way of interaction without blame. Employing active listening skills ensures that you take time to listen to people without interrupting or blaming them.

4. *POSITIVITY AND OPPORTUNISM: Transform any view of failure into an opportunity to improve. Focus on the best scenario rather than what is not good enough.* If you are putting yourself, others, or the invention/solution down and if you keep thinking things are never going to change, it is likely to have a negative effect. Practicing a mindset of confidence, positivity, and having a "can do" attitude is more effective and can help change a hitherto negative view. It is easy to define a minimum baseline of what is absolutely needed, and what is not good enough; but is more important to add a vision of what is the best possible scenario. Enthusiasm and envisioning the optimal "keep the bar high."

5. *COMPASSION AND EMPATHY: Make your fears clear and provide solutions to mitigate them. Accept and respect others' fears and know that you do not need to fully understand them to show empathy and compassion for yourself and others.* If there is something you are uncomfortable with, it is important to talk about what would remove the feeling. In this way, the optimal invention/solution scenario seen from your point of view is provided. At the same time, you need to be able to listen to others' points of view and to hear what their best options look like. A new, improved level of trust may be established when you are able to share such sentiments; this naturally has great prospects.

6. *WILLINGNESS TO LET GO: Understand that a new, improved invention/solution may require termination of a current solution or change of a current invention.* Very often it is essential to let go of an old way of doing things in order to implement a new one. Not everybody is ready for this, and there can be many reasons. It is important to understand the optimal timing for a change, but also that improvement may challenge the existing and require a certain level of sacrifice and adjustment.

7. *PATIENCE: Be willing to wait for people to gain understanding and for the time to be right for change.* It takes time to gather people and necessary resources to support a new invention or solution. Knowing this may help you become aware of where in the process you find yourself and help you accept that change takes time.

4.3   OVERVIEW OF STEPS I - IX

Each of the steps I – IX elucidates aspects that must be addressed in order to create sustainable inventions/solutions:

In the section following Table 4.1 (Steps I – IX: Questions Section) I pose a set of questions that will create a basis for an action plan when answered. A summary of the purpose and outcomes from each step precedes these questions. Both the requirements for sustainable innovation and the attitudes earlier described are included. Please note that these questions relate to addressing opposition, the inclusion of unlikely combinations, and the development of a "plan B," and they imply that the essential resources and technology are or will be available.

Table 4.1:

Step I: Identification of innovation potential and recipients of innovation	An examination of whether or not the invention/solution meets an essential need is performed. All stakeholders are identified (all people who are affected by the invention/solution).
Step II: Justification of innovation	The requirements for sustainable innovation are addressed (as defined by the checklist in Chapter 2).
Step III: Examination of innovation barriers ("mining for gold")	If there are major obstacles to progress, they are specified. These could include a number of practical issues and also emotional reactions such as anger, fear, excitement, cultural values, ideology, tradition, etc. Identification of how anger and fear can be used to add value to the final "product" is included here.
Step IV: Building a vision	The best possible scenarios seen by the different stakeholders are determined. Each person describes his or her own vision and specific goals. This section is used to identify all aspects and "extremes" of individual wishes.
Step V: Applying integrative intelligence	So-called "integrative intelligence" is applied. It revisits all of the sustainable innovation attitudes mentioned in this chapter. The goal is to identify overlaps and congruencies between Steps III and IV. Most importantly, it provides the largest possible (and often conditional) expansion of the overlap or congruencies (by virtue of compromise/flexibility) and lists tangible outcomes.
Step VI: Prioritization of outcomes and establishing measures of success	Prioritization of tangible outcomes is performed. The definitions of measures of success for each outcome are determined.
Step VII: Adjustment	An adjustment based on these measures (or lack of these) may be necessary and is based on the anticipation of variation of outcomes.
Continues.	

	Table 4.1:
Continued.	
Step VIII: Establishing a back-up plan	The final cornerstone of sustainable innovation, the back-up plan is established. This step ensures that in case of failure, disaster is avoided. Said in popular terms, "emergency fire doors" are implemented.
Step IX: Limitations	Limitations to innovation/solution-making are identified and evaluated. This step can sometimes be performed prior to any of the other steps, but it often requires that a thorough analysis (such as steps I-VIII has taken place). It includes assessing the subset of personalities and the level of resistance to change. A particular focus is on lack of ability or willingness to employ emotional intelligence, strong ideals, lack of common interests, and timing.

4.4   STEPS I – IX: QUESTIONS SECTION

STEP I: IDENTIFICATION OF INNOVATION POTENTIAL AND RECIPIENTS OF INNOVATION

Purpose:

1. To define the essential need.

2. To identify all parties who may have different or opposing views about the inventions/solutions and the outcome of these or simply constitute an unusual combination of backgrounds and knowledge/ideas. These parties must be present, represented, or consulted during the process of innovation/solution-making.

Outcomes:

1. Specific invention/solution parameters.

2. Participant list.

Questions:

1. What is the problem/issue/need?

2. Why is the problem/issue/need deemed essential?

3. How is the problem/issue/need manifesting itself?

4. How would a invention/solution improve the situation?

5. Who is currently affected by the problem/issue/need?

6. Who/what would benefit from the invention/solution?

7. Who/what would be adversely affected by the invention/solution?

8. Who would remain unaffected by the invention/solution?

9. How would matching of unlikely combinations ("plurosis") be ensured?

STEP II: JUSTIFICATION OF INNOVATION

Purpose:

1. To ensure responsibility, superiority, and peace.

2. To triage invention/solution if the criteria of responsibility, superiority, and peacefulness are not met.

Outcomes:

1. Definition of the specific parameters required for sustainable innovation as identified by the checklist in Chapter 3.

2. Decision regarding initiation or termination of project.

Questions:

1. Has the need for an essential invention/solution been identified (see Step I)?

2. Define how the invention/solution can be considered honorable, truthful, and trustworthy.

3. How does the invention/solution show an interest in the common good?

4. Who would be personally responsible for the invention/solution?

5. How would the best intentions for the affected parties be ensured?

6. How would the best intentions for the affected parties be demonstrated?

7. What is the substantial positive impact to beneficiaries?

8. How would the beneficial impact of the invention/solution be measured?

9. What is an acceptable environmental impact?

10. How would an adverse impact of the invention/solution be measured?

11. Would the invention/solution consume reasonable and acceptable amounts of resources?

12. Would the invention/solution produce acceptable, re-usable, or no waste?

13. Would the invention/solution include one or more back-up plans in the form of a substitute innovation?

14. Would the invention/solution include one or more back-up plans in the form of the development of independent self-sustainable units?

15. How would superiority be measured?

16. How would unusual combinations of ideas, materials, methods or people be ensured?

17. How would the synergistic or additive effect of composition of ideas, materials, methods or people be measured?

18. How would opposition and other concerns be addressed and included in the final invention/solution?

19. How would the ability to find "common ground" be ensured and measured?

STEP III: EXAMINATION OF INNOVATION BARRIERS ("MINING FOR GOLD")

Purpose:

1. "Mining for gold" by identifying of the reasons for anger, fear, opposition, and understanding of alternative views.

2. To promote a "ripe environment" and "ripe time" for the invention/solution.

Outcomes:

1. Definition of what the overall "opposition landscape" (defined by anger, fear, and non-emotional opposition) and alternative suggestions look like.

2. Definition of which aspects can make a invention/solution better (by addressing the "opposition landscape" and alternative views).

3. Specifications regarding the need for emotional or physical changes prior to introduction of a invention/solution.

Questions:

1. Are any of the stakeholders angered by the proposed invention/solution?

2. If yes, what are the reasons for the anger?

3. How can changes be introduced to the invention/solution so that anger is diffused?

4. Do any of the stakeholders express fears regarding the proposed invention/solution?

5. If yes, what are these fears?

6. How can changes be introduced to the invention/solution so that fear is mitigated?

7. Are any of the stakeholders opposed to the proposed invention/solution?

8. If yes, what are the reasons (other than anger and fear, such as old habits, institutional practices, unclear rules or regulations, etc.) for this opposition?

9. How can changes be introduced to the invention/solution so that the opposition stops?

10. Are additional changes needed to address physical conditions for the stakeholders?

11. If yes, please specify (food, water, clothing, shelter, safety, healing, etc.).

12. Are additional changes needed to address the emotional state of the stakeholders?

13. If yes, please specify (counseling, forgiveness, etc.).

14. Are any of the stakeholders simply providing alternative views to the invention/solution?

15. If yes, what are these specific views?

16. How can changes be introduced to the invention/solution so that these views strengthen the final product?

STEP IV: BUILDING A VISION

Purpose:

1. To identify individual perspectives, visions and specific wishes.

Outcomes:

1. Specifications of the optimal scenario as seen through an individual's eyes.

2. A list of all of the optimal outcomes.

Questions:

1. What are the stakeholder-specific wishes and visions that determine the minimally-accepted positive outcome of the invention/solution?

2. What are the stakeholder's specific wishes and visions that determine the optimally-positive outcome of the invention/solution?

STEP V: APPLYING INTEGRATIVE INTELLIGENCE

Purpose:

1. To identify congruencies and overlap of aspects and aspirations for the invention/solution by integrating knowledge from Steps III and IV.

2. To expand these as much as possible (identify the most inclusive situation/define the highest level of "plurosis").

Outcomes:

1. An improved invention/solution based on the integration of new aspects and aspirations as identified by individual stakeholders.

2. A time-line for introduction of the invention/solution.

3. A list of the tangible outcomes and action items.

Notes:

This step is the most challenging and requires application of all of the sustainable innovation attitudes described in Chapter 4. The reader is encouraged to re-examine his or her mindset before performing this step:

1. INTENT AND CLARITY: Agree that there is something to improve/invent/resolve, clearly define it, and strive to achieve it.

2. FLEXIBILITY AND CURIOSITY: Be flexible, display openness, assume that others' input is important for the end solution, and that you do not have all the answers yourself. Understand that society is dynamic and that you need to be flexible too.

3. FORGIVENESS AND RECONCILIATION: Avoid focusing on what has been fair and unfair; instead focus on a new best and just outcome. Avoid blaming anybody; instead display a forgiving or reconciliatory mindset. Use active listening skills.

4. POSITIVITY AND OPPORTUNISM: Transform any view of failure into an opportunity to improve. Focus on the best scenario rather than what is not good enough.

5. COMPASSION AND EMPATHY: Make your fears clear and provide solutions to mitigate them. Accept and respect others' fears and know that you do not need to fully understand them to show empathy and compassion for yourself and others.

6. WILLINGNESS TO LET GO: Understand that a new improved invention/solution may require termination of a current solution or change of a current invention.

7. PATIENCE: Be willing to wait for people to gain understanding and for the time to be right for change.

Questions:

1. Determine how outcomes from Step III can be integrated into the innovation/solution by answering the following:

    (a) The aspects that by integration can make a invention/solution better (by addressing the anger) without substantially changing the invention/solution are:

    (b) The aspects that by integration can make a invention/solution better (by addressing the fear) without substantially changing the invention/solution are:

    (c) The aspects that by integration can make a invention/solution better (by addressing the non-emotional opposition) without substantially changing the invention/solution are:

    (d) The aspects that by integration can make a invention/solution better (by addressing the alternative views) without substantially changing the invention/solution are:

    (e) The physical changes that can promote a "ripening effect" and can be implemented prior to the introduction of a invention/solution without substantially changing the invention/solution are:

    (f) A realistic timeline* for this "ripening effect" is:

    Please see Step IX to learn when a ripening effect cannot be expected and it is time to let go of the attempt to provide innovation or change

    (g) The emotional changes that can promote a "ripening effect" and can be implemented prior to the introduction of a invention/solution without substantially changing the invention/solution are:

    (h) A realistic timeline for this "ripening effect" is:

2. Determine how the outcomes from Step IV can be integrated into the invention/solution by answering the following:

    (a) The specifications of the minimally-acceptable scenarios as seen (the smallest amount of flexibility/compromise), which can be implemented without reinvigorating any anger, fear, non-emotional opposing views, or eliminate valuable alternative views are:

    (b) The optimal outcomes as seen with individual eyes (the largest amount of flexibility/compromise) that can be implemented without reinvigorating any anger, fear, non-emotional opposing views, or eliminate valuable alternative views are:

3. A realistic timeline for implementation that is not going to reinvigorate any anger, fear, non-emotional opposing views, or eliminate valuable alternative views and takes into account the "ripening effect" for the necessary physical and emotional changes is:

4. The tangible outcomes and corresponding action items are:

STEP VI: PRIORITIZATION OF OUTCOMES AND ESTABLISHING MEASURES OF SUCCESS

Purpose:

1. To prioritize the tangible outcomes for the most inclusive situation.

2. To define the measures of success for each tangible outcome.

Outcomes:

1. A priority list of the outcomes.

2. A definition of how success is measured for each outcome.

Questions:

1. What is the priority of each tangible outcome?

2. How will success be measured in each case?

STEP VII: ADJUSTMENT

Purpose:

1. To anticipate the level of variation in the outcomes.

2. To plan adjustments to the outcomes and action items based on these variations.

Outcomes:

1. A description of the variations and necessary adjustments.

Questions:

1. What type and extent of variation can be anticipated for each outcome?

2. What corresponding adjustments can be made to each outcome and measures of success for each outcome?

STEP VIII: ESTABLISHING A BACK-UP PLAN

Purpose:

1. To provide one or more back-up plans for the innovation/solution.

Outcomes:

1. An alternative to the innovation/solution in case of failure due to method.

2. An alternative to the innovation/solution in case of failure due to size or location.

Questions:

1. If the invention/solution fails due to method, which kinds of replacement methods or processes will be immediately available?

2. If the invention/solution fails due to size or location, which kinds of replacement methods or processes will be immediately available?

STEP IX: LIMITATIONS

Purpose:

1. To identify situations in which invention/solution-making is not possible or is severely restricted.

Outcomes:

1. An evaluation form that can be used to make an informed decision regarding whether or not to pursue the invention/solution.

Questions:

1. Which of the following scenarios are true?

    (a) Steps I – VIII are not adequately described.

    (b) Essential resources and/or technologies are not available.

    (c) Existing inventions/solutions are considered "good enough."

    (d) Involved parties (decision makers/inventors/solution-makers) are dependent on the continuation of existing solutions. Reasons could be monetary, practical, political, or other.

    (e) Involved parties (decision makers/inventors/solution-makers) are resistant to investing in updates or better inventions/solutions. Reasons could be monetary, practical, political, or other.

    (f) Ripening effect is severely delayed or not happening.

    (g) Involved parties (decision makers/inventors/solution-makers) display no congruencies or overlaps between opinions/ideas.

    (h) Involved parties (decision makers/inventors/solution-makers) are suffering from a lack of ability or willingness to employ emotional intelligence.

If "true" is the answer to any of the statements, it is possible that the invention/solution should be reconsidered. It may not be possible to achieve a sustainable situation.

CHAPTER 5

Notes on the Application to 21st Century Challenges

This chapter provides notes pertinent to answering the questions outlined in Chapter 4. Four areas in which we face severe challenges are addressed: energy production and consumption, food supply, emerging diseases, and verbal disputes. The need for innovation/solution-making in each area is briefly described (part of step 1) as are selected steps of the method. A more detailed exploration is necessary to create discrete inventions/solutions; this is beyond the scope of this chapter. Following the examples is a personal step-by-step work section for developing your own sustainable ideas and inventions (see Chapter 6).

5.1 SUSTAINABLE ENERGY PRODUCTION AND CONSUMPTION (STEPS I – III)

STEP I: IDENTIFICATION OF INNOVATION POTENTIAL AND RECIPIENTS OF INNOVATION

Scenario: Today's challenges include the inadequate amount and uneven distribution of energy resources. In addition, current energy supply methods seem to contribute to climate changes through the introduction of pollution and over-utilization of natural resources. These issues are a function of both higher demand for energy per capita and population growth. Without any changes to current supply methods or a reduction in demand, some people may be without power to meet their essential needs for food production, temperature adjustment, and transportation, and the amount of inhabitable land may decrease due to rising sea levels and pollution. The need for improvement is obviated when power outages occur or when energy prices rise. How can new and inexhaustible energy resources be found, used to produce a greater amount of energy, and be better distributed while reducing pollution? How can the energy need be lowered? People from many fields may contribute to invent an optimal solution: biologists, engineers, climate change specialists, environmentalists, architects, geologists, astronomers, meteorologists, supply chain specialists, professionals from industries with very high power consumption, and others.

STEP II: JUSTIFICATION OF INNOVATION

A sustainable invention/solution must serve the common good. This could mean that a cumulative energy need will be met at a reasonable cost, and that according to the best of our current knowledge, the environment will not be harmed as defined by specific measures and thresholds. Considerations also relate to the consumption of energy. If Earth's limited natural resources are to be ideally utilized, then they need to be "stretched" as much as possible, which means effectiveness must be optimized. Alternatively, if inexhaustible resources, such as solar rays, wind, waves and other naturally-occurring movements can be used, this is of lesser concern. Perhaps the energy from the occurrence of weather storms could be exploited? Regardless of the chosen energy source, backup plans are needed. In case of failure of or large variation in one type of supply, such as oil, an automatic supply from another source needs to be readily available. It is advisable to include more than one backup plan for very sensitive operations (such as electronic transactions). Establishing the "emergency fire door" principle could mean that failure of an operation would be localized and that alternative ways of continuing operations would always be in place.

STEP III: EXAMINATION OF INNOVATION BARRIERS ("MINING FOR GOLD")

Opposing views are "golden" when seen from an innovation perspective. The best invention or solution is one that recognizes these views as valid and takes into account how to minimize or halt opposition. In regards to energy consumption, oil is a limited resource. Oil has worked very well for many years; there has been a great enough readily-available supply to meet demand and its production has been a lucrative endeavor. These favorable conditions have somewhat stifled the search for similarly profitable energy resources. Perhaps the old saying of, "if it works, then don't fix it" has been the dominant thinking platform and it has precluded developing a "plan B." Persuasive reasoning, such as identification of new effective solutions or actual urgency (exemplified in the meltdown of nuclear plants), for an alternative is essential for change, and it often goes hand-in-hand with some level of compromise or time of transition.

In general, if opposition to a new method is caused by others simply having better ideas, this is a very promising situation because an improvement is likely to be provided directly. Contrary, if opposition is caused by a history of scandals or unfairness, assurances of future ethical conduct may be needed before any change can be reasonably anticipated. Most importantly, if opposition to the invention/solution is based on the anticipation of lowered future living standards or a lesser quality of essentials, such as food, water, clothing, shelter, safety, and healing, discussion of how to avoid such negative impacts is certain to improve the invention/solution.

5.2 ENVIRONMENTALLY-FRIENDLY, ABUNDANT FOOD PRODUCTION (STEPS III – V)

Scenario: At least three issues come to mind immediately when it comes to food supply and food distribution systems. Many people go to bed hungry both in the so-called underdeveloped and developed countries. Others have enough food or too much of it, and it goes to waste. Some food is produced in ways that pollute the Earth and cause a variety of health concerns for both humans and crop pollinators. How can this be altered so that each person would have the right amount of healthy food to eat that was produced in an environmentally-friendly way? Many people can contribute to help answer this question: nutritionists, doctors, farmers, food and crop scientists, supply chain managers, climate experts, religious experts, and others.

STEP III: EXAMINATION OF INNOVATION BARRIERS ("MINING FOR GOLD")

There are several potential "gold-mines to dig" in the food supply area. Some of these relate to food production and others to the level of sharing and modes of food distribution. One school of thought is that what we have done so far using conventional farming is working well and does not need change. Another view is that to feed the growing world population, large-scale food production using conventional farming methods and gene modified crops is necessary. Yet a third view is that farming without pesticides or organic farming practices are best. Finally, some are of the opinion that specific farming practices are less important than effective distribution ensuring enough food for everybody, and that food is not wasted. Let us take a closer look at the opportunities buried in these views.

People who oppose organic farming often feel negatively impacted by this practice in some way, or they believe that organic produce is neither better nor healthier than conventionally-produced or genetically modified produce. In addition, there is often the opinion that organic farming is not able to feed enough people to be taken seriously. Opponents of conventional farming and gene-modified crop production are worried that large food producers may develop monopolies on certain crops. These crops often require pesticides that have to be purchased from the very same food producers that provide the seed or plants. Pesticides are thought to pollute soil and ground water. There is concern that gene modified seeds and progeny cannot be contained, and that these will spread to other crops and undomesticated plants. The fear is that within a short while, all crops will depend on pesticides to grow to maturity, which leads to a higher risk of pest resistance to the chemicals used, to increased pollution, and to negative impacts on a variety of species. In addition to these concerns, there is an overriding thought that simply eating foods exposed to pesticides and/or gene modified crops is not safe. In summary, the opposition is manifold and, therefore, so are the opportunities for improvement.

STEP IV: BUILDING A VISION

If the minimally-accepted outcome includes the acceptance of all farming practices, then all parties can do what they think is best based on their specific goals and intentions. This is only realistic if there is no impact of one method on another and as long as all concerns have been addressed to each parties' satisfaction. Specifically, what can be done to encompass the variety of viewpoints? It seems like a data check including information about environmental pollution, dispersal of gene-modified crops, and food safety as well as the cost of the different practices is needed. In addition, establishing a base-line of how much food is needed is a pre-condition for any sustainable invention/solution. Perhaps surrogate digestive systems could be used to study what happens when we eat food produced in different ways? Perhaps such studies could be the new "gold standard" for testing anything edible? Last, food distribution systems and how to avoid wasting food must be considered and whether or not it is responsible, practical, and cost-effective to transport food over very long distances, except in situations of emergency.

STEP V: APPLYING INTEGRATIVE INTELLIGENCE

Once data provides the necessary facts, these can be used to strengthen food production and distribution practices. If there is no need to worry about pollution, dispersal, monopoly, or safety, the anger or fear caused by these worries can be eliminated or sufficiently reduced. Alternatively, if there are still concerns, great caution must be applied until the unwanted impacts are eliminated or tolerated. Once this situation occurs, it may be apparent which method(s) are preferable in which locations. Perhaps the scenario of mixing different practices to achieve an optimal outcome is possible. This could mean taking advantage of the best of all methods; avoid pollution and other safety hazards, while using technology in a manner deemed responsible. Even though concerns have been addressed to satisfaction, there may still be doubts regarding whether or not to change. Additional data and time may be needed to provide a necessary "ripening effect."

5.3   CONTROL OF EMERGING DISEASES (STEPS VI – VIII)

Scenario: The occurrence of infectious diseases to which we, livestock, and crops are susceptible can be devastating. There are many reasons to anticipate the coming or increase of such pests. Monoculture and reduced species and genetic diversity put "evolutionary pressure" on pathogens and are likely to make an outbreak more severe. Increased trade and mobility enhance the speed with which an infection can spread. An unhealthy lifestyle further decreases chances of survival. As with the other examples in this chapter, this issue is highly complex, and an invention/solution can benefit from the input of multiple parties such as: medical doctors, epidemiologists, microbiologists, farmers, food producers, agronomists, veterinarians, transportation specialists, pilots, engineers, architects, ministers, psychologists.

STEP VI: PRIORITIZATION OF OUTCOMES AND ESTABLISHING MEASURES OF SUCCESS

Let us assume that by using steps I – IV, parties have agreed to inventions/solutions that result in the following hypothetical outcomes:

1. A certain level of diversity is maintained among people, animals, and crops within a habitat (the definition of this habitat is measured by its size). A successful outcome means that diversity has been defined in terms of numbers of a certain species and how many different species are present within the habitat.

2. Trade of goods and mobility of people can be traced to the source immediately. A measure of success is real-time establishment of the origin and destination of shipments and travelers and an ability to isolate a geographic area and population effectively.

3. All people take a minimum level of responsibility for their health. A successful outcome means healthy weight and no substance dependencies (unless medically prescribed).

4. New diseases are anticipated and medicines are developed. This means medicines must be available and able to reach relevant populations at the right times.

   Some outcomes are more realistic or likely than others, and it is important to discuss what matters most to prioritize outcomes. Also, variation can be anticipated for each outcome, and corresponding adjustments need to be planned (step VII) but are not shown here.

STEP VIII: ESTABLISHING A BACK-UP PLAN

It is easy to understand the importance of a "plan B" in the case of emerging diseases. If there is a severe uncontrollable outbreak, what is the strategy to survive? What would an "emergency fire door" look like? It may mean quarantine of an area, new eradication or sterilization regimens, or perhaps personal isolation suits. It may mean new types of medicines. Another "plan B" could be that of an individual having reflected upon how to prepare in a dignified manner to not survive.

5.4 VERBAL DISPUTES AND CONSTRUCTIVE FREEDOM OF SPEECH (STEPS IV – V, IX)

Scenario: There are many examples of unnecessarily hurtful things that have been said. It happens at all levels, between children, teens, adults, and between people of different cultures, backgrounds, and convictions. Problems arise when people feel threatened in regard to differing opinions. This can be aggravated when they also have differing views of what is acceptable to say and whether or not self-control should be applied to a situation. Some people think it is ideologically important to say anything they want without edit or contemplating the consequences. Others have a strict code of conduct in regard to which kinds of topics or expressions are allowed or not allowed. If these two types speak together, discontent is likely simply based on different communication choices, in addition to any disagreement caused by differing opinions. Sometimes disagreements are expressed with frustration, resentment, anger, and even threats and violence. Fortunately, many people exhibit diplomacy in dealing with conflict. These people would prefer that everybody get along, and they agree that freedom of speech is important, but that what we say has consequences and should be considered responsibly. These people can help "bridge the gap" between the, often much louder, disputing groups, and people of all three opinions (and differing degrees of these) must be involved if a sustainable solution is to be reached.

STEP IV: BUILDING A VISION

Let us assume that steps I – III have revealed that there is a polar difference in opinions surrounding a certain matter. "Mining for gold" usually shows that each point of view carries value and opportunity to invent a better situation than the existing one. There is often truth in criticism. The question becomes how enough "bravery" can be mustered by an individual to embrace the criticism and extract the truth. Unfortunately, the minimally-accepted vision is often hard to comprehend. There really is not a "let live" option because one wants something the other doesn't want and vice versa. Because the interaction between the two exacerbates the issue, neither one of the sides can get exactly what is desired if the parties are going to be in touch with each other and live peacefully. So the options are: to fight it out until one wins followed by coercing the other into submission (this does not convince or convert the coerced party), "draw a line in the sand" and stop interacting, or redefine the dispute to something acceptable for most people.

STEP V: APPLYING INTEGRATIVE INTELLIGENCE

Polarized debates create opportunity for social, ideological, and other types of invention. One could say it is a situation of unlikely combinations of opinions. Adding responsibility and a peaceful purpose to the mindset is necessary. "Sifting the gold" from step III in order to establish what will work for both parties needs to be done. Which issues cause the anger, the fear, or the non-emotional opposition? How can it be made clear that these issues have been understood and how can they be addressed so a invention/solution can be optimized? Is there reason to believe more time is needed for the process to be successful, or can specific actions be taken to accelerate the process? Each party must define the specifics of the minimally- and maximally-acceptable scenarios. The discussion could be about health care, immigration, taxes, energy, communication styles, or any other topic. A common code of communication conduct would be helpful to avoid wasting time and resources and to promote the sustainable innovation paradigm. Such a code would require self-control of the communicators and contain elements of clear and to-the-point speech. They would avoid loaded emotional elements of irony, satire, or typical communication errors, such as irrelevant comparisons, which can easily be perceived as "put downs."

STEP IX: LIMITATIONS

There will likely always be people who consistently oppose most others for numerous reasons, such as visibility, fame, inflexibility, or strong ideals, and this may cause incongruencies to be found. Other challenges exist when people are unable to exert verbal or physical self-control or when they see criticism as a personal attack. Another limitation is apparent when people express apathy or if they generally fear change. Becoming aware of these limitations is as important as identifying the opportunities.

CHAPTER 6

Personal Step-by-Step Work Section

STEP I: IDENTIFICATION OF INNOVATION POTENTIAL AND RECIPIENTS OF INNOVATION

Purpose:

1. To define the essential need.

2. To identify all parties who may have different or opposing views about the inventions/solutions and the outcome of these or simply constitute an unusual combination of backgrounds and knowledge/ideas. These parties must be present, represented, or consulted during the process of innovation/solution-making.

Outcomes:

1. Specific invention/solution parameters.

2. Participant list.

Questions:

1. What is the problem/issue/need?
   ....
   ....
   ....

2. Why is the problem/issue/need deemed essential?
   ....
   ....
   ....

3. How is the problem/issue/need manifesting itself?
   ....
   ....
   ....

4. How would a invention/solution improve the situation?

....
....
....

5. Who is currently affected by the problem/issue/need?

....
....
....

6. Who/what would benefit from the invention/solution?

....
....
....

7. Who/what would be adversely affected by the invention/solution?

....
....
....

8. Who would remain unaffected by the invention/solution?

....
....
....

9. How would matching of unlikely combinations ("plurosis") be ensured?

....
....
....

Revisiting purpose:

- To define the essential need.

- To identify all parties who may have different or opposing views about the inventions/solutions and the outcome of these or simply constitute an unusual combination of backgrounds and knowledge/ideas. These parties must be present, represented or consulted during the process of innovation/solution-making.

Revisiting outcomes:

- Specific invention/solution parameters.

- Participant list.

STEP II: JUSTIFICATION OF INNOVATION

Purpose:

1. To ensure responsibility, superiority, and peace.

2. To triage the invention/solution if the criteria of responsibility, superiority, and peacefulness are not met.

Outcomes:

1. Definition of the specific parameters required for sustainable innovation as identified by the checklist in Chapter 3.

2. Decision regarding initiation or termination of project.

Questions:

1. Has the need for an essential invention/solution been identified (see Step I)?

   ....
   ....
   ....

2. Define how the invention/solution can be considered honorable, truthful, and trustworthy.

   ....
   ....
   ....

3. How does the invention/solution show an interest in the common good?

   ....
   ....
   ....

4. Who would be personally responsible for the invention/solution?

   ....
   ....
   ....

5. How would the best intentions for the affected parties be ensured?

   ....
   ....
   ....

6. How would the best intentions for the affected parties be demonstrated?

   ....
   ....
   ....

7. What is the substantial positive impact to beneficiaries?

....
....
....

8. How would the beneficial impact of the invention/solution be measured?

....
....
....

9. What is an acceptable environmental impact?

....
....
....

10. How would an adverse impact of the invention/solution be measured?

....
....
....

11. Would the invention/solution consume reasonable and acceptable amounts of resources?

....
....
....

12. Would the invention/solution produce acceptable, re-usable, or no waste?

....
....
....

13. Would the invention/solution include one or more back-up plans in the form of a substitute innovation?

....
....
....

14. Would the invention/solution include one or more back-up plans in the form of the development of independent self-sustainable units?

....
....
....

15. How would superiority be measured?

      ....
      ....
      ....

16. How would unusual combinations of ideas, materials, methods, or people be ensured?

      ....
      ....
      ....

17. How would the synergistic or additive effect of composition of ideas, materials, methods or people be measured?

      ....
      ....
      ....

18. How would opposition and other concerns be addressed and included in the final invention/solution?

      ....
      ....
      ....

19. How would the ability to find "common ground" be ensured and measured?

      ....
      ....
      ....

Revisiting purpose:

- To ensure responsibility, superiority, and peace.

- To triage the invention/solution if the criteria of responsibility, superiority and peacefulness are not met.

Revisiting outcomes:

- Definition of the specific parameters required for sustainable innovation as identified by the checklist in Chapter 3.

- Decision regarding initiation or termination of project.

STEP III: EXAMINATION OF INNOVATION BARRIERS ("MINING FOR GOLD")

Purpose:

1. "Mining for gold" by identifying the reasons for anger, fear, opposition, and understanding of alternative views.

2. To promote a "ripe environment" and "ripe time" for the invention/solution.

Outcomes:

1. Definition of what the overall "opposition landscape" (defined by anger, fear, and non-emotional opposition) and alternative suggestions look like.

2. Definition of which aspects can make a invention/solution better (by addressing the "opposition landscape" and alternative views).

3. Specifications regarding the need for emotional or physical changes prior to introduction of a invention/solution.

Questions:

1. Are any of the stakeholders angered by the proposed invention/solution?
   . . . .

2. If yes, what are the reasons for the anger?
   . . . .
   . . . .
   . . . .
   . . . .
   . . . .
   . . . .
   . . . .
   . . . .
   . . . .
   . . . .
   . . . .

3. How can changes be introduced to the invention/solution so that anger is diffused?
   . . . .
   . . . .
   . . . .
   . . . .

....

....

....

....

....

....

....

....

4. Do any of the stakeholders express fears regarding the proposed invention/solution?

....

5. If yes, what are these fears?

....

....

....

....

....

....

....

....

....

....

....

....

6. How can changes be introduced to the invention/solution so that fear is mitigated?

....

....

....

....

....

....

....

....

....

....

....

7. Are any of the stakeholders opposed to the proposed invention/solution?

....

8. If yes, what are the reasons (other than anger and fear, such as old habits, institutional practices, unclear rules or regulations, etc.) for this opposition?

....

....

....

....

....

....

....

....

....

....

....

....

9. How can changes be introduced to the invention/solution so that the opposition stops?

....

....

....

....

....

....

....

....

....

....

....

....

10. Are additional changes needed to address physical conditions for the stakeholders?

....

11. If yes, please specify (food, water, clothing, shelter, safety, healing, etc.).

....

....

....

....

....

....

....

....
....
....
....

12. Are additional changes needed to address the emotional state of the stakeholders?
....

13. If yes, please specify (counseling, forgiveness, etc.).
....
....
....
....
....
....
....
....
....
....
....
....

14. Are any of the stakeholders simply providing alternative views to the invention/solution?
....

15. If yes, what are these specific views?
....
....
....
....
....
....
....
....
....
....
....
....

16. How can changes be introduced to the invention/solution so that these views strengthen the final product?

....

....

....

....

....

....

....

....

....

....

....

Revisiting purpose:

- "Mining for gold" by identification the reasons for anger, fears, and opposition, and understanding of alternative views.

- To promote a "ripe environment" and "ripe time."

Revisiting outcomes:

- Definition of what the overall "opposition landscape" (defined by anger, fear, and non-emotional opposition) and alternative suggestions look like.

- Definition of which aspects can make a invention/solution better (by addressing the "opposition landscape" and alternative views).

- Specifications regarding the need for emotional or physical changes prior to the introduction of a invention/solution.

STEP IV: BUILDING A VISION

Purpose:

1. To identify individual perspectives, visions and specific wishes.

Outcomes:

1. Specifications of the optimal scenario as seen through an individual's eyes.

2. A list of all of the optimal outcomes.

Questions:

1. What are the stakeholder-specific wishes and visions that determine the minimally-accepted positive outcome of the invention/solution?

....

....

....

....

....

....

....

....

....

....

....

....

2. What are the stakeholder's specific wishes and visions that determine the optimally-positive outcome of the invention/solution?

....

....

....

....

....

....

....

....

....

....

....

....

Revisiting purpose:

- To identify individual perspectives, visions and specific wishes.

Revisiting outcomes:

- Specifications of the minimally acceptable scenario as seen with individual eyes.

- A list of all of the optimal outcomes.

STEP V: APPLYING INTEGRATIVE INTELLIGENCE

Purpose:

1. To identify congruencies and overlap of aspects and aspirations for the invention/solution by integrating knowledge from Steps III and IV.

2. To expand these as much as possible (identify the most inclusive situation/define the highest level of "plurosis").

Outcomes:

1. An improved invention/solution based on the integration of new aspects and aspirations as identified by individual stakeholders.

2. A time-line for introduction of the invention/solution.

3. A list of the tangible outcomes and action items.

Notes:

This step is the most challenging and requires application of all of the sustainable innovation attitudes described in Chapter 4. The reader is encouraged to re-examine his or her mindset before performing this step:

1. INTENT AND CLARITY: Agree that there is something to improve/invent/resolve, clearly define it, and strive to achieve it.

2. FLEXIBILITY AND CURIOSITY: Be flexible, display openness, assume that others' input is important for the end solution, and that you do not have all the answers yourself. Understand that society is dynamic and that you need to be flexible too.

3. FORGIVENESS AND RECONCILIATION: Avoid focusing on what has been fair and unfair; instead focus on a new best and just outcome. Avoid blaming anybody; instead display a forgiving or reconciliatory mindset. Use active listening skills.

4. POSITIVITY AND OPPORTUNISM: Transform any view of failure into an opportunity to improve. Focus on the best scenario rather than what is not good enough.

5. COMPASSION AND EMPATHY: Make your fears clear and provide solutions to mitigate them. Accept and respect others' fears and know that you do not need to fully understand them to show empathy and compassion for yourself and others.

6. WILLINGNESS TO LET GO: Understand that a new improved invention/solution may require termination of a current solution or change of a current invention.

7. PATIENCE: Be willing to wait for people to gain understanding and for the time to be right for change.

Questions:

1. Determine how outcomes from Step III can be integrated into the invention/solution by answering the following:

   (a) The aspects that by integration can make a invention/solution better (by addressing the anger) without substantially changing the invention/solution are:

   . . . .
   . . . .
   . . . .
   . . . .
   . . . .
   . . . .

   (b) The aspects that by integration can make a invention/solution better (by addressing the fear) without substantially changing the invention/solution are:

   . . . .
   . . . .
   . . . .
   . . . .
   . . . .
   . . . .

   (c) The aspects that by integration can make a invention/solution better (by addressing the non-emotional opposition) without substantially changing the invention/solution are:

   . . . .
   . . . .
   . . . .
   . . . .
   . . . .
   . . . .

   (d) The aspects that by integration can make a invention/solution better (by addressing the alternative views) without substantially changing the invention/solution are:

   . . . .
   . . . .
   . . . .
   . . . .
   . . . .
   . . . .

   (e) The physical changes that can promote a "ripening effect" and can be implemented prior to the introduction of a invention/solution without substantially changing the invention/solution are:

....
....
....
....
....
....

(f) A realistic timeline* for this "ripening effect" is:

....
....

*Please see Step IX to learn when a ripening effect cannot be expected and it is time to let go of
the attempt to provide innovation or change.

(g) The emotional changes that can promote a "ripening effect" and can be implemented
prior to the introduction of a invention/solution without substantially changing the in-
vention/solution are:

....
....
....
....
....
....

(h) A realistic timeline for this "ripening effect" is:

....
....

2. Determine how the outcomes from Step IV can be integrated into the invention/solution by
answering the following:

(a) The specifications of the minimally-acceptable scenarios as seen with individual eyes (the
smallest amount of flexibility/compromise), which can be implemented without reinvig-
orating any anger, fear, non-emotional opposing views, or eliminate valuable alternative
views are:

....
....
....
....
....
....
....
....
....

. . . .

. . . .

. . . .

(b) The optimal outcomes as seen with individual eyes (the largest amount of flexibility/compromise) that can be implemented without reinvigorate any anger, fear, non-emotional opposing views, or eliminate valuable alternative views are:

. . . .

. . . .

. . . .

. . . .

. . . .

. . . .

. . . .

. . . .

. . . .

. . . .

. . . .

3. A realistic timeline for implementation that is not going to reinvigorate any anger, fear, non-emotional opposing views, or eliminate valuable alternative views and takes into account the "ripening effect" for the necessary physical and emotional changes is:

. . . .

. . . .

. . . .

. . . .

. . . .

. . . .

4. The tangible outcomes and corresponding action items are:

Outcome	Action item

Revisited purpose:

- To identify congruencies and overlap of aspects of and aspirations for the invention/solution by integration of knowledge from Steps III and IV.

- To expand these as much as possible (identify the most inclusive situation = define the highest level of "plurosis").

Revisited outcomes:

- An improved invention/solution based on the integration of new aspects and aspirations as identified by individual stakeholders.

- A time-line for introduction of the invention/solution.

- A list of the tangible outcomes and action items.

STEP VI: PRIORITIZATION OF OUTCOMES AND ESTABLISHING MEASURES OF SUCCESS

Purpose:

1. To prioritize the tangible outcomes for the most inclusive situation.

2. To define the measures of success for each tangible outcome.

Outcomes:

1. A priority list of the outcomes.

2. A definition of how success is measured for each outcome.

Questions:

1. What is the priority of each tangible outcome?

2. How will success be measured in each case?

Priority	Outcome	Measures of success
1		
2		
3		
4		
5		
6		
7		
8		
9		

Revisited purpose:

- To prioritize the tangible outcomes.
- To define the measures of success for each tangible outcome.

Revisited outcomes:

- A priority list of the outcomes.
- A definition of how success is assessed for each outcome.

STEP VII: ADJUSTMENT

Purpose:

1. To anticipate the level of variation in the outcomes.
2. To plan adjustments to the outcomes and action items based on these variations.

Outcomes:

1. A description of the variations and necessary adjustments.

Questions:

1. What type and extent of variation can be anticipated for each outcome?
2. What corresponding adjustments can be made to each outcome and measures of success for each outcome?

Variation	Adjusted outcome	Adjusted measures of success

Revisited purpose:

- To anticipate the level of variation in the outcomes.

• To plan adjustments to the action items based on these variations.

Revisited outcomes:

• A description of the variations and necessary adjustments.

STEP VIII: ESTABLISHING A BACK-UP PLAN

Purpose:

1. To provide one or more back-up plans for the innovation/solution.

Outcomes:

1. An alternative to the innovation/solution in case of failure due to method.

2. An alternative to the innovation/solution in case of failure due to size or location.

Questions:

1. If the invention/solution fails due to method, which kinds of replacement methods or processes will be immediately available?

   ....
   ....
   ....
   ....
   ....
   ....
   ....
   ....
   ....

2. If the invention/solution fails due to size or location, which kinds of replacement methods or processes will be immediately available?

   ....
   ....
   ....
   ....
   ....
   ....
   ....
   ....
   ....

Revisited purpose:

- To provide one or more back-up plans for the innovation/solution.

Revisited outcomes:

- An alternative to the innovation/solution in case of failure due to method.

- An alternative to the innovation/solution in case of failure due to size or location.

STEP IX: LIMITATIONS

Purpose:

1. To identify situations in which invention/solution-making is not possible or is severely restricted.

Outcomes:

1. An evaluation form that can be used to make an informed decision regarding whether or not to pursue the invention/solution.

Questions:

1. Which of the following scenarios are true?

Scenario	True	False
Steps I – VIII are not adequately described.		
Essential resources and/or technologies are not available		
Existing inventions/solutions are considered "good enough."		
Involved parties (decision makers/inventors/solution-makers) are dependent on the continuation of existing solutions. Reasons could be monetary, practical, political, or other.		
Involved parties (decision makers/inventors/solution-makers) are resistant to investing in updates or better inventions/solutions. Reasons could be monetary, practical, political, or other.		
Ripening effect is severely delayed or not happening.		
Involved parties (decision makers/inventors/solution-makers) display no congruencies or overlaps between opinions/ideas.		
Involved parties (decision makers/inventors/solution-makers) are suffering from a lack of ability or willingness to employ emotional intelligence.		

If "true" is the answer to any of the statements, it is possible that the invention/solution should be reconsidered. It may not be possible to achieve a sustainable situation.

Revisited purpose:

- To identify situations in which invention/solution-making is not possible or very restricted.

Revisited outcomes:

- An evaluation form which can be used to make an informed decision regarding whether or not to pursue the invention/solution.

CHAPTER 7

Looking to the Future

7.1 IMPACT OF SUSTAINABLE INNOVATION

UNDERSTANDING VALUE SYSTEMS

Consciously bringing unlikely materials, methods, and people together means that information from different sources is going to mingle to a large extent and through different channels. This already occurs when people travel to different areas for leisure or to study and conduct business in new places. Cultures are continuously mixed through migration and relations, in schools, at work, through social media, and in families. There is increased focus on interdisciplinary interaction. Although differences in values and value systems are bound to create initial conflict, the sustainable innovation regimen presented in this book teaches that this is often followed by opportunities for improvement. This can take place through the creation of an enhanced "hybrid situation" based on increased communication and understanding. The exposure to and tolerance of other ways of thinking and doing things that this mixing causes has several implications. It promotes a growing level of openness towards others and a willingness to learn why other thoughts are conceived and different actions are performed.

POSITIVE CURIOSITY AND RESPECTFUL DISAGREEMENT

While it is often challenging to come to agreements, it frequently seems that it is even harder for us to disagree peacefully. Letting go of fear of the unfamiliar and the unchallenged protectionism of the "status quo" and instead replacing these with positive curiosity is a very promising and brave position to take. Those who do so will be able to enjoy the benefits of learning how the existing can be improved or positively altered. Curiosity creates the foundation for an environment ready for sustainable innovation. At the same time, it is a primer for communication that ensures enhanced understanding. Most importantly, maintaining a positive curiosity and communication ensures that a healthy level of respect for disagreement can be achieved, and that we are becoming genuinely interested in each others' inventions and solutions. When parties do not agree to an invention, common solution, or way of living, diversity and the fact that there is more than one way of doing things may still be celebrated.

VIRTUOUS PLANNING AND ENRICHED SOLUTIONS

Virtuous conduct/ethical behavior is rarely addressed when inventions are planned. In other words, this is an untapped innovation resource. It is often ignored because virtue and character may have associations such as "duty," "extra work that doesn't pay off," "unnecessary charity," or simply "not relevant to the task at hand." It is my hope that by now you may have decided that good and strong ethical conduct is an integral part of sustainable innovation. Bringing together many different ideas, materials, methods, and people ensures that we use a combinatorial approach for innovation. One very exciting aspect is an unexpected "plurosis" effect that may take place when we least expect it. Finally, the important ideal of "doing no harm" also applies to behavior when plans fail. Sustainable innovation includes adapting to emergencies and planning for contingencies.

7.2   CONSEQUENCES OF GLOBAL COMPETITION

We are used to thinking about how to best compete with others to become the #1 country or person at something (for example innovation, wealth, and power). This has long benefited us because it means people strive to always do well. It works well as a motivator as long as someone else does not work as effectively and as long as there is a power and wealth differential. But times are changing. Many populations are working harder and more intensely to reach these pinnacles. This means competition is continuously increasing and everything has to be more cost effective than before. Jobs may be given to those who will work the fastest for the lowest wages, production takes place where it is cheapest, and the quality of services and products is decreasing as a result. Competing in this way means a never-ending reduction in quality of living for all. There is not even hope it will stop at the lowest possible denominator as long as the getting more or the most for the money and out-innovating and out-competing each other is our mindset because the quality bar will constantly be lowered. Instead, if we can convert from a sole focus on competition to a drive for sustainable innovation regardless of who is #1, perhaps many #1s in many countries in many fields may be discovered, and collectively we will be able to celebrate the abundance of resources at our disposal.

Today, there is no one set of conditions that fit all people of all countries. Globalization clearly has its effect, and there is a high level of mixing of populations, cultures and knowledge, which ultimately will lead to more similarities in the future. The question becomes whether or not we want the same conditions for all. Some may consider this a very good and fair solution. Others may think it will stifle competition, progress, and innovation, and if we all have the same conditions, there would be fewer incentives to strive for anything. Also, being the same may eliminate the much-needed diversity of thinking, which is so essential to progress. How do we harvest the benefits of friendly competition or the joy of innovation while securing a reasonable standard of living for all? The proposed sustainable solution is a compromise (and the spirit of this book).

7.3   A GLOBAL WIN-WIN SCENARIO

We could strive for a certain set of minimal living conditions for all. This would provide a base-line standard of living. At the same time, it is important to nurture innovation and preserve freedom for uniqueness in interdependent, self-sustainable countries, regions, and local businesses. Each of us must work responsibly to contribute what is needed to achieve this ideal vision, while removing ourselves from the thinking that we are entitled to basic or increasing wealth without contributing to the common good. It may seem like going back in time, but it is merely re-learning the basic ways of survival, outreach, and progress shown by our forefathers, while additionally and concurrently taking advantage of the best that today's societies and innovation can provide. Engaging government, educational leaders, commerce, industry, and the general public in the discussion will ensure that the much-needed variation is represented.

If we value sustainable innovation both globally and locally, this will help to preserve resources and diversity. Global entities may elect to create smaller units that could satisfy special local needs and create the "emergency fire doors." Local areas may revisit taking care of their own basic needs (for example, food and energy), depending on a few central services (for example, higher education and advanced healthcare), and exchanging unique goods and services within their own and between other areas and, if wanted, globally. A region could build and maintain its own expectations and balance in terms of standard of living, services, exchange, and innovation. Such a model requires that people within local areas have a wide range of skills, are content with what the region can provide, can reach an agreement about their expectations, that they support the local economy, that they make room for local entrepreneurs, and they accept global solutions to global problems. This may minimize disparities within regions and mitigate the threat of systemic failure. It may help us focus our attention on what is essential instead of superfluous because we would have to produce most of what we need locally. We may elect to depend on local and regional trade and rely much less on foreign trade, which in this way can be placed in its correct context - a welcome asset when it is present but not essential to a region's survival (unless a situation of emergency has occurred).

In the end we must ask ourselves, "Were we content with the way we spent our time on Earth? Did we choose to confirm our place in the animal kingdom by subscription to survival of the fittest in its most primitive sense, preoccupied with getting the most from the smorgasbord of wealth, power, and entertainment? Or did we share our resources and inventions and work virtuously for a life in unity, collaboratively belonging to and caring for each other and the Earth? Did we invent to the best of our abilities, for the common good, with gratitude for all of what we have been given?" If you are reading this, there is still time. Thank you for trying.

Inspirational Reading[1]

[1] Borbye et al.: Industry Immersion Learning: Real-Life Case Studies in Biotechnology and Business. Wiley Blackwell (2009). 1

[2] Atkisson, A. et al: WorldChanging: A User's Guide for the 21st Century (ed. Steffen, A.). Abrams (2008). 3, 4, 5

[3] Tapscott, D. & Williams, A. D.: Wikinomics, How Mass Collaboration Changes Everything. Portfolio (2006). 4

[4] Mezrich, B.: The Accidental Billionaires: The Founding of Facebook: A Tale of Sex, Money, Genius and Betrayal. Doubleday (2009). 4

[5] Sacks, J.: The Dignity of Difference. How to Avoid the Clash of Civilizations. Continuum (2003). 4

[6] Kelley, T: The Ten Faces of Innovation. IDEO'S strategies for Beating the Devil's Advocate & Driving Creativity throughout Your Organization. Currency Doubleday (2005). 4

[7] Pink, D.H.: A Whole New Mind. Why Right-Brainers will Rule the Future. Riverhead Books (2005). 5

[8] MacKenzie, G.: Orbiting the Giant Hairball. A Corporate Fool's Guide to Surviving with Grace. Viking Penguin (1998). 5

[9] Gergen, C. & Vanourek, G.: Life Entrepreneurs. Ordinary People Creating Extraordinary Lives. Jossey Bass (2008). 5

[10] Fredrickson, B. L.: Positivity. Three Rivers Press (2009). 5, 11

[11] Borbye, L.: Out of the Comfort Zone: New Ways to Teach, Learn, and Assess Essential Professional Skills. Morgan & Claypool Publishers (2010). 11

[1]Please note that these are not necessarily representing primary references but inspired this work.

[12] Pink, D. H.: Drive. The Surprising Truth About What Motivates Us. Riverhead Books (2009). 11

[13] Borbye, L.: Secrets to Success in Industry Careers: Essential Skills for Science and Business. Academic Press (2008). 11, 16

[14] Lipton, B.L.: The Biology of Belief. Hay House (2009). 11

[15] Rosenthal, N. E.: The Emotional Revolution. Citadel Press (2002). 12

[16] Rosenberg, M. B.: Nonviolent Communication. Puddle Dancer Press (2003). 13

[17] McDonough, W. & Braungart, M.: Cradle to Cradle: Remaking the Way We Make Things. North Point Press (2002). 4

[1]Please note that these are not necessarily representing primary references but inspired this work.

APPENDIX A

Notes on Teaching Sustainable Innovation

Challenges. It is of great value to understand how environment inspires or discourages innovative thinking. Three issues are particularly important:

1. When people are brought up with excessive entertainment provided by television or by other means or otherwise lack "space to ponder", they may experience difficulty imagining new inventions.

2. When public display of a person's abilities is promoted and praised uncritically this may in the long term cause a focus on self-promotion, difficulty receiving critique and reluctance toward working in teams.

3. Many people spend a large part of their days receiving information and, often, experience an "information overload" This can create a tendency to quickly "believe" and "copy and paste" easily-accessible information and result in a lack of depth and authenticity.

Overcoming challenges. Knowing that there are issues in regard to innovative thinking, teamwork, self-awareness, depth and authenticity, how can sustainable innovation be taught and "boosted?" The following list is inspired by [11] and gives a variety of tips to teachers, and in some cases these can be extrapolated to entrepreneurs and leaders of small or large enterprises:

1. Create a need for innovation. This can be done by requiring innovation, include teamwork, authenticity, etc. for a top grade, a bonus, and/or a promotion.

2. Create a specific challenge or a problem to solve. This gives people a starting point.

3. Remove resources (time, money, processes, specific items, etc). Lack of resources will make people "stretch themselves" to find a solution.

4. Change the parameters often. This could mean removal or change of resources (such as time-lines, funding, format of deliverable, etc.).

5. Think "big." Require that people imagine the best-possible scenarios.

6. Inspire. Bring external entrepreneurs and inventors to the table.

7. Provide minimal assistance. Establish a network of mentors who are capable of guiding innovation without revealing an invention/solution.

8. Ensure that opposing views will be present, unlikely combinations can occur, and that "plan B's" are developed (see Chapters 1, 2, and 3).

9. Create multi-disciplinary teams of people with diverse backgrounds (see Appendix B).

10. Develop training regimens for sustainable innovation using the method described in Chapter 4.

APPENDIX B

Examples of Unlikely Combinations Parameters

(The context can be found in Chapters 1, 2, and 3).

- Age

- Sex

- Ethnicity

- Country of origin

- Culture

- Ideological orientation

- Religious conviction

- Spiritual beliefs

- Education level

- Field of education

- Other fields of expertise

- Marital status

- Level of global understanding

- "Life experiences" (such as child birth, marriage, divorce, injury, disease, death, etc.)

- Social economic standard

- Personality type

- Environmental preference

- Inclusion or subtraction of different components may relate to a number of essential, structural or other differences such as product specifics (origin, composition, design, etc.), viability measures (productivity, longevity, etc.), and indication or usage.

APPENDIX C

Sustainable Innovation Requirements Checklist

(The context can be found in Chapter 3).

- *NEED: Meet essential need(s) or improve a situation, and provide substantial positive impact to beneficiaries.* Parties must agree on what the term positive impact means and identify how it can be measured.

- *VIRTUOUS CONDUCT: Be honorable (as defined by actions to be proud of based on an established common ethical code), truthful (telling the absolute truth and providing the necessary details), trustworthy (upholding the truth and ethical code of conduct), and show interest in the common good* defined by a positive impact on the majority of affected parties.

- *ACCOUNTABILITY: Assign personal accountability and display best intentions for affected parties.* Inventors/solution-makers must be identifiable and held responsible for their actions/inventions. The intentions should be clearly expressed.

- *CONSUMPTION: Create reasonable, acceptable consumption, produce only acceptable, re-usable, or no waste, and cause acceptable impact on the environment.* The parameters for how many resources can be used to create a certain invention/solution and how much waste it will likely create should be established. The parameters also include how much and which kinds of waste can be tolerated, if any.

- *ALTERNATIVES: Include one or more back-up plans (substitute innovation or breakdown into independent units).* Establishing these plans makes us think about the end-products' flexibility. We must practice anticipating the future best and worst case scenarios, which can contribute to substantial value-adding and avoid or mitigate unexpected disasters.

- *SUPERIORITY: Have clear measures of superiority.* Knowing what it means to create a superior solution includes knowing how to measure whether it was achieved. This includes establishing tangible measures of success.

- *UNLIKELY COMBINATIONS: Use unusual combinations of ideas, materials, methods or people, or provide enrichment by including opposition and concerns.* Providing evidence for a mutual, heterosis, or "plurosis" effect is important. Without taking this effect into account, it is very likely that another, more superior innovation is easily conceivable.

- *COMPROMISE: Be able to find "common ground."* A prerequisite for inclusion of different ideas, materials, methods, opinions, or people is that flexibility will be displayed and/or compromise will take place.

APPENDIX D

Sustainable Innovation Attitudes Checklist

(The context can be found in Chapter 4).

- *INTENT AND CLARITY: Agree that there is something to improve/invent/resolve, clearly define it, and strive for a invention/solution.* Setting a goal of what to achieve is instrumental to success. Without this goal, it is unclear what needs to be done, and the likelihood of a change is low. It is important to want to strive to meet the goal. Some conflicting attitudes are those in which you A) do not care about whether or not a invention/solution will be found and B) do not want a invention/solution for any number of reasons.

- *FLEXIBILITY AND CURIOSITY: Be flexible, display openness, assume that others' input is important for the end solution, and that you do not have all the answers yourself. Understand that society is dynamic and that you need to be flexible too.* When discussing and including other peoples' opinions and ideas, the invention/solution is strengthened substantially. Being able to "borrow" a different set of brains is a gift, and being able to borrow many simultaneously could create a revelation! Try not to be intimidated by criticism or great additions to your own ideas; instead find the value. Society is likely to change faster than you do, and to keep up, you will need the help of others.

- *FORGIVENESS AND RECONCILIATION: Avoid focusing on what has been fair and unfair; instead focus on a new best and just outcome. Avoid blaming anybody; instead display a forgiving or reconciliatory mindset. Use active listening skills* [13]. What has happened in the past can easily create preconceptions regarding the future. Much has likely been less than optimal, but it is not going to help to keep rehashing it. Without naïveté, try to look objectively at inventions/solutions and dare to think the best. A conversation in which people are made to feel guilty is rarely productive. There are other ways to address issues that you do not want to have happen again. Say exactly what outcomes are desired and what you want from people rather than what you do not want or what others have done wrong. Also, when you are able to forgive what has happened in the past, you may feel free to pursue a new way of interaction without blame. Employing active listening skills ensures that you take time to listen to people without interrupting or blaming them.

- *POSITIVITY AND OPPORTUNISM: Transform any view of failure into an opportunity to improve. Focus on the best scenario rather than what is not good enough.* If you are putting yourself, others, or the invention/solution down and if you keep thinking things are never going to change, it is likely to have a negative effect. Practicing a mindset of confidence, positivity, and having a "can do" attitude is more effective and can help change a hitherto negative view. It is easy to define a minimum baseline of what is absolutely needed, and what is not good enough; but is more important to add a vision of what is the best possible scenario. Enthusiasm and envisioning the optimal "keep the bar high."

• *COMPASSION AND EMPATHY: Make your fears clear and provide solutions to mitigate them. Accept and respect others' fears and know that you do not need to fully understand them to show empathy and compassion for yourself and others.* If there is something you are uncomfortable with, it is important to talk about what would remove the feeling. In this way, the optimal invention/solution scenario seen from your point of view is provided. At the same time, you need to be able to listen to others' points of view and to hear what their best options look like. A new, improved level of trust may be established when you are able to share such sentiments; this naturally has great prospects.

• *WILLINGNESS TO LET GO: Understand that a new, improved invention/solution may require termination of a current solution or change of a current invention.* Very often it is essential to let go of an old way of doing things in order to implement a new one. Not everybody is ready for this, and there can be many reasons. It is important to understand the optimal timing for a change, but also that improvement may challenge the existing and require a certain level of sacrifice and adjustment.

• *PATIENCE: Be willing to wait for people to gain understanding and for the time to be right for change.* It takes time to gather people and necessary resources to support a new invention or solution. Knowing this may help you become aware of where in the process you find yourself and help you accept that change takes time.

APPENDIX E

Examples of Drivers and Barriers to Sustainable Innovation

(The context can be found in Chapter 1).

Common Drivers

- Monetary gains
- Intellectual property laws
- Basic human needs
- Idealism
- Enthusiasm
- Pleasure
- Play
- Exploration

Common Barriers

- Monetary gains
- Intellectual property laws
- Basic human needs
- Ethnic, spiritual, ideological or practical differences
- Lack of collaborative communication
- Lack of monetary and structural resources
- Lack of technologies and logical processes

- Existing inventions/solutions are considered "good enough"

- Dependency on the continuation of existing solutions

- Resistance to investing in updates or better inventions/solutions.

- Timing

- Lack of congruencies or overlaps between opinions/ideas

- Lack of ability or willingness to employ emotional intelligence

Other Books by the Author

Secrets to Success in Industry Careers: Essential Skills for Science and Business.

By L. Borbye
Academic Press. ISBN 9780123738691

People are often surprised by the differences between the environments of the university and industry work when adjusting to their first jobs in industry. Such adjustment often takes 9 – 12 months and is costly for the employer in terms of lost effectiveness. Potential employees can improve their employment readiness and competitiveness by learning about the new environment prior to entry. *Secrets to Success in Industry Careers: Essential Skills for Science and Business* introduces the reader to the differences between what is needed in school and what is needed in industry. It describes how to select a job and the entire process of obtaining a job including analysis of a job description, writing an application, preparation for an interview and conduct during and after an interview. Most importantly, this book is the ideal "industry-insider" guide because it provides an overview of skills and understanding essential for success on the job. Topics include: business goals and bottom line, leadership and entrepreneurship, communication skills, marketing, discipline, flexibility, creativity and out-of-the-box thinking, ambiguity management, intellectual property, specialty technology and knowledge, quality, ethics, globalization, expectation management, and career management. It is highly relevant for people who are seeking jobs in industry, already employed in industry, employing others in industry, teaching others about what it is like to work in industry, or people who are simply curious about the industry environment and culture. The secret is that mastering these skills gives success both on the job and in life.

Key Features:

- Topics give well-rounded and "to the-point" instruction for mastery of a wealth of technical and managerial skills.

- Fictional anecdotes make it easy to understand application and context.

- Description of career selection and expectation management helps the reader make effective choices throughout his/her career.

- Self assessments tools give means to establish the reader's strengths and opportunities for growth.

- Summary of key skills gives a quick way of reviewing whether the reader is on the track to success.

Table of Contents:

Preface

PART I: Career Choice Considerations and Job Pursuit

1. Comparison of Academic and Industry Environments
2. Choosing an Industry Career
3. Professional Conduct During the Job Application Process

PART II: Professional Conduct on the Job

4. Business Goals and Bottom Line
5. Leadership and Teamwork
6. Communication Skills
7. Marketing
8. Discipline
9. Flexibility
10. Creativity and Out-of-the-Box Thinking
11. Ambiguity Management
12. Intellectual Property
13. Specialty Technology and Knowledge
14. Quality
15. Ethics
16. Globalization
17. Expectation Management
18. Career Management

Epilogue

Index

Industry Immersion Learning: Real-Life Case Studies in Biotechnology and Business.

By L. Borbye et al.
Wiley Blackwell. ISBN 9783527324088

Few textbooks are available that contain hands-on training for students aiming to work in industry and instructions for university teachers interested in collaborating with industry. This book is ideal for professional science Master's and MBA students, as well as university teachers and industry professionals. It provides guidance on how to approach industry professionals and create educational alliances. The strategy of establishing contact to industry employers and the process of developing case studies are outlined. Eight re-usable real-life case studies including a description of the topic, mission and goals, learning outcomes, study plan, student deliverables, and teacher's instructional materials are provided by industry professionals from the Research Triangle Park in North Carolina. Among the case studies are examples of how to identify biomarkers and new drugs simultaneously, prioritize and develop products in compliance with rules and regulations, commercialize products and protect and manage intellectual property, optimize processes and technologies for manufacturing, and minimize human errors in production.

Key Features:

- Provides detailed instruction for professors and other teachers about how to contact industry professionals and how to develop educational projects in collaboration.

- Presents model case studies meant to inspire interdisciplinary solutions, innovative thinking and professional skills application.

- Invites teachers and students to learn from existing case-studies in a range of topics from discovery and commercialization to intellectual property management and best production practices.

Table of Contents:

Preface

1. Principles of Industry Immersion Learning
2. Integration of Pharmaceutical and Diagnostic Co-Development and Commercialization: Adding Value to Therapeutics by Applying Biomarkers
3. Pharmaceutical Industry Product Portfolio Planning and Management
4. Entrepreneurship: Establishing a New Biotechnology Venture
5. Introduction to U.S. Patent Law

6. Intellectual Property Management

7. Excellence In Pharmaceutical Manufacturing

8. Aligning Behaviors and Standards in a Regulated Industry: Design and Implementation of a Job Observation Program

Index

Out of the Comfort Zone: New Ways to Teach, Learn, and Assess Essential Professional Skills.

By L. Borbye
Morgan & Claypool Publishers. ISBN 9781608451753 and ISBN 9781608451760 (ebook)

This book describes a simple, common sense method of how to include professional skills training in any curriculum without compromising academic rigor. It relies on introduction of unanticipated, yet manageable crises simulating scenarios commonly experienced in the workplace. Examples include: how to respond to a demand for innovation and teamwork, a lay-off, a re-organization, or switching jobs and projects. Preparing and practicing a mindful and healthy response is beneficial, and it serves as a platform for attitude training and character building prior to unexpected events on the job and elsewhere. Also included are student reflections on learning and a rubric to assess the professional skills learning outcomes.

Key Features:

1. Description of the importance of, incentives for, and rewards of exiting the comfort zone

2. Principles for teaching and learning professional skills

3. Student anecdotes and reflections

4. Rubric entries and assessment of learning

Table of Contents:

Preface

1. The Comfort Zone and "Being out of It"

2. Exiting the Comfort Zone: Reasons and Impact

3. Getting Educators and Students out of the Comfort Zone

4. Principles of "Out-of-the-Comfort-Zone" (OOC) Teaching

5. Anecdotes of OOC Learning

6. Measuring the Outcome

Bibliography

Author's Biography

Author's Biography

LISBETH BORBYE

Lisbeth Borbye, Ph.D., is a pioneer in alliance building between industry and universities and development of innovative, need-based instructional materials for students and professors.

Dr. Borbye holds the positions of Assistant Dean for Professional Education at North Carolina State University and Director of the University of North Carolina's System wide Professional Science Master's (PSM) Initiative. PSM programs seek to meet the needs for an improved graduate workforce in industry, non-profit organizations and government by providing interdisciplinary career-specific education with both depth and breadth. In addition, these degree programs bring important cultural changes to universities through collaboration with employers and inclusion of professional skills such as basic business skills, teamwork, leadership, effective communication, innovative thinking and entrepreneurship. Dr. Borbye is chairing university-wide, system-wide and national PSM advisory boards. She is the author of the books entitled "Secrets to Success in Industry Careers: Essential Skills for Science and Business" (Academic Press), and "Out of the Comfort Zone: New Ways to Teach, Learn, and Assess Essential Professional Skills" (Morgan & Claypool Publishers), and co-author of "Industry Immersion Learning: Real-Life Case Studies in Biotechnology and Business."

Printed in the United States
by Baker & Taylor Publisher Services